I0471959

FEDERAL EMERGENCY MANAGEMENT AGENCY
UNITED STATES ADMINISTRATION
NATIONAL FIRE DATA CENTER

Profile of the Urban Fire Problem in the United States

Prepared by:

TriData Corporation
1000 Wilson Boulevard
Arlington, Virginia 22209

This publication was produced under Contract EMW-95-C-4717 by TriData Corporation for the United States Fire Administration, Federal Emergency Management Agency. Any information, findings, conclusions, or recommendations expressed in this publication do not necessarily reflect the views of the Federal Emergency Management Agency or the United States Fire Administration.

May 1999

TABLE OF CONTENTS

Executive Summary

This report characterizes the nature of the fire problem in urban areas of the United States. Urbanized areas have large populations, and they typically have higher densities of people and buildings than rural areas. Publications are available that characterize the overall U.S. fire problem and the fire problem in rural areas, but there has not been a recent profile of fire in urban areas.[1] This report addresses that need.

Using 1996 data on urban fires, this study found that:

- Fires that occurred outdoors were the most common type of urban fire reported. In 1996, 42 percent of urban fires were classified as outdoor fires, just under one-third were structure and vehicle fires, and two percent of fires occurred in "other" locations.

- The leading cause of outdoor fires in urban areas was incendiary or suspicious origin.

- While outdoor fires were most numerous, structure fires accounted for the vast majority of fire deaths, fire injuries, and property loss associated with urban fires.

- *Non-residential Structures:* Fires of incendiary or suspicious origin predominated among non-residential structure fires, accounting for 30 percent of fires.

- *Residential Structures:* Cooking fires accounted for over one-quarter of all home fires. Incendiary and suspicious origin ranked second, followed by heating and electrical distribution.

- The leading causes of residential fires were relatively consistent throughout the four major regions of the country. In every region (Northeast, Midwest, South, and West) cooking fires were the leading cause. Incendiary or suspicious origin was the second leading cause in every region but the Northeast, where heating fires were second.

---

[1] U.S. Fire Administration, 1998, *Fire in the United States: 1986-1995,* Tenth Edition, Publication FA-183/August 1998, (Emmitsburg, MD: U.S. Fire Administration) and U.S. Fire Administration, 1998, *The Rural Fire Problem in the United States,* Publication FA-180/August 1998, (Emmitsburg, MD: U.S. Fire Administration).

- The leading causes of fatal residential fires were also relatively consistent across the country. Smoking was the leading cause of fatal home fires in every region except the West, where fires of incendiary or suspicious origin ranked first.

- A higher proportion of residential structure fires occurred in apartments in urban areas compared to the U.S. as a whole. This is likely due to the fact that more of the urban housing stock is comprised of multifamily housing.

- The prevalence of apartments in the urban housing stock may also account for the lesser role of heating fires. Heating fires in rural areas are often associated with chimneys and woodstoves, or other alternate heating devices. Most apartments have central heating only, reducing the risk of fires associated with alternate heating.

- A majority (54 percent) of urban home fires occurred where no working smoke detectors were present. Similarly, 69 percent of fires with one or more fatalities occurred in homes not protected by operating detectors. These rates are similar to, though slightly lower than, rates for the U.S. as a whole.

Introduction

This report characterizes the nature of the fire problem in urban areas of the United States. Urbanized areas have large populations, and they typically have higher densities of people and buildings than rural areas. These differences make the fire problem in urban areas worthy of separate study. For example, research has shown that the rate of structure fires due to incendiary or suspicious origin is a greater problem in communities with populations of 100,000 or more than it is in smaller towns and rural areas.[2] On the other hand, heating fires are typically less of a problem in urban areas than rural areas.[3] This report uses a subset of 1996 data from the National Fire Incident Reporting System to draw a general profile of urban fires.

While it is important to profile urban fires, it is also interesting to look for variations within this category. Differences in climate and building stock across regions could lead to slightly different urban fire profiles. This study investigates regional differences in the causes of residential structure fires. For example, it is likely that home fires related to heating occur more frequently in northern areas of the U.S. Similarly, electrical distribution fires are likely to be more common in the Northeast and South, where the housing stocks are older on average than in areas of the Midwest and West. Regional differences in fire death rates are also presented. These rates were calculated based on mortality data from the National Center for Health Statistics.

After discussing urban fires generally, nine different metropolitan areas in the U.S. are profiled separately. These areas are Atlanta, Boston, Chicago, Cleveland, Dallas, Hartford, Houston, San Francisco, and Washington-Baltimore. Fire death rates for each of these areas are also presented. Taken together, the profiles of these metropolitan areas show how the urban fire problem can vary from place to place.

Where the Data Came From

Two sources of data are relied upon in this report. The first is fire incident and casualty data from the National Fire Incident Reporting System (NFIRS). NFIRS is a data system maintained by the U.S. Fire Administration. It was established in the late

---

[2] National Fire Protection Association, 1997, *U.S. Arson Trends and Patterns – 1996*, (Quincy, MA: NFPA), pp. 18-19.

[3] U.S. Fire Administration, 1998, *The Rural Fire Problem in the United States*, Publication FA-180/August 1998, (Emmitsburg, MD: U.S. Fire Administration), pp. 11-12.

1970s, and today it is the largest fire data set in the United States. Each year almost one million new fires are added to NFIRS. Annually, fire departments from all over the country report on the number and types of fires to which they have responded. The system is voluntary, but it is estimated that close to half the nation's fire departments participate in NFIRS and almost half of all fires attended by fire departments are reported.

For the purposes of this study, "urban" fires were defined as those occurring in a select group of metropolitan areas. These areas were chosen based on three criteria: large population size, geographic location, and NFIRS participation. Table 1 provides a list of the 18 metropolitan areas included in the data set.

Table 1. Metropolitan Areas Included in the "Urban" Fire Data Set

Metropolitan Area	Metropolitan Area
Atlanta	Kansas City
Boston-Worcester-Lawrence	Los Angeles-Riverside-Orange County
Chicago-Gary-Kenosha	Miami-Fort Lauderdale
Cincinnati-Hamilton	Milwaukee-Racine
Cleveland-Akron	Minneapolis-St. Paul
Dallas-Ft. Worth	New York-Northern New Jersey-Long Island
Denver-Boulder-Greeley	San Diego
Hartford	San Francisco-Oakland-San Jose
Houston-Galveston-Brazoria	Washington-Baltimore

Each of the metropolitan areas listed above is a census-designated Metropolitan Statistical Area (MSA) or a Consolidated Metropolitan Statistical Area (CMSA). The raw "urban" data set including all these MSAs consisted of 324,571 fire incidents. These fires represent 38 percent of all fires reported to NFIRS in 1996. To further narrow the definition of "urban," only incidents reported by fire departments located in central counties of metropolitan areas were included in the analysis of the general urban fire problem.[4] The designation of central counties was taken from the U.S. Department of Agriculture's Rural-Urban Continuum. The final urban data set contained 306,775 fire incidents.

---

[4] These counties are coded "0" by the Department of Agriculture. They are central counties of metropolitan areas with populations of one million or more.

The second source of data used in this report is mortality data from the National Center for Health Statistics. The mortality data used for this study spanned from 1991 to 1995. An overall average fire death rate for all the urban areas included in this study is presented, as well as individual rates for each of the metropolitan areas profiled at the end of the report. Average fire death rates are presented because, particularly at the metropolitan level, annual numbers of fire deaths can vary significantly from year to year. Fires tend to be somewhat random events, though their frequency can be impacted by factors such as the severity of a winter or other weather conditions. The 1995 mortality data were the most recent data available at the time of this study, so the death rates presented are the average of five years of data, from 1991 to 1995.

The mortality data were aggregated based on the county in which the fire death occurred. These county-level data were then aggregated into metropolitan areas using 1996 census definitions. To ascertain an overall "urban" fire death rate for the U.S., only those counties with populations of one million or more were included in the analysis.

A Profile of Urban Fires

This section presents an overview of urban fires that were reported to NFIRS in 1996. To begin, the types of situations that fire departments responded to are presented, and then the leading causes of these fires are identified. The discussion then focuses on structure fires and residential structure fires in particular. Structure fires account for the majority of deaths, injuries, and dollar loss associated with fire in the U.S. An analysis of the presence and operability of smoke alarms in residential structure fires is also presented.

Distribution and Cause of Urban Fires

The majority of urban fires occur outdoors. Figure 1 shows that 42 percent of urban fires occurred outdoors. Structure and vehicle fires each accounted for less than one-third of urban fire incidents reported to NFIRS in 1996, and "other" fires made up the remaining 2 percent.

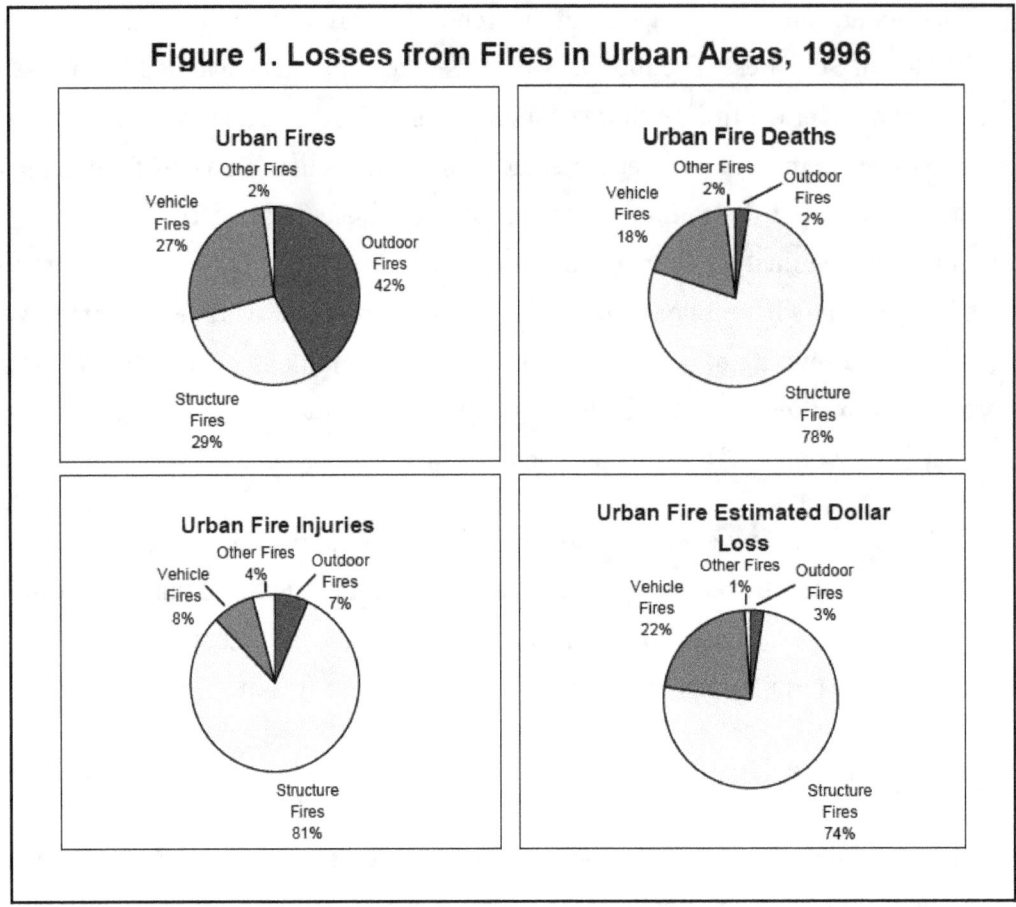

Figure 1. Losses from Fires in Urban Areas, 1996

Urban Fires
- Other Fires 2%
- Vehicle Fires 27%
- Outdoor Fires 42%
- Structure Fires 29%

Urban Fire Deaths
- Other Fires 2%
- Outdoor Fires 2%
- Vehicle Fires 18%
- Structure Fires 78%

Urban Fire Injuries
- Other Fires 4%
- Outdoor Fires 7%
- Vehicle Fires 8%
- Structure Fires 81%

Urban Fire Estimated Dollar Loss
- Other Fires 1%
- Outdoor Fires 3%
- Vehicle Fires 22%
- Structure Fires 74%

Source: 1996 NFIRS data

While outdoor fires are the most common type of urban fire in the U.S., Figure 1 shows that structure fires account for the majority of fire losses, both in human and property terms. According to 1996 NFIRS data, urban structure fires accounted for at least three-fourths of all urban fire deaths, injuries, and estimated dollar loss.

Figures 2 and 3 show outdoor and vehicle fires broken down by cause of fire. In both cases, incendiary or suspicious origin was the leading cause of fire. This is a cause for concern. In the case of outdoor fires, one concern is that some juvenile firesetters use outdoor fires as "gateway" fires. They move from setting fires outdoors to setting fires in vehicles or structures. Arson in structures increases the likelihood that someone will be injured or killed as a consequence of the fire, whether intended or not.[5]

---

[5] U.S. Fire Administration, 1997, *Arson in the United States*, Publication FA-174/August 1997, (Emmitsburg, MD: U.S. Fire Administration), p. 3.

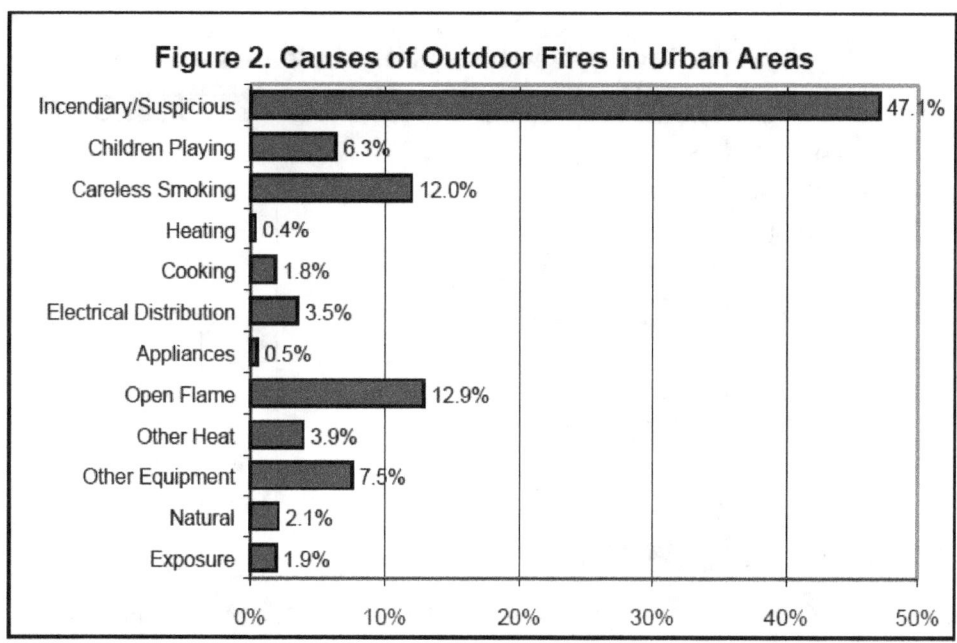

Source: 1996 NFIRS data

In the case of vehicle arson, all Americans who drive pay the cost of these fires in the form of increased insurance premiums. Insurance fraud is an important motivation in vehicle arson. Another problem is the negative impact vehicle arson has on neighborhoods. This dynamic is exemplified by the case of the Dudley neighborhood in Boston. Medoff and Sklar report that during the 1980s, Dudley, a low-income, predominantly African American neighborhood, served as a dumping ground for cars abandoned by their owners or stolen and stripped of parts. These cars were often burned. As a result of these activities, the residents of the neighborhood paid a high cost in terms of a diminished quality of living in the area.[6]

---

[6] Peter Medoff and Holly Sklar, 1994, *Streets of Hope: The Fall and Rise of an Urban Neighborhood* (Boston, MA: South End Press), p. 73.

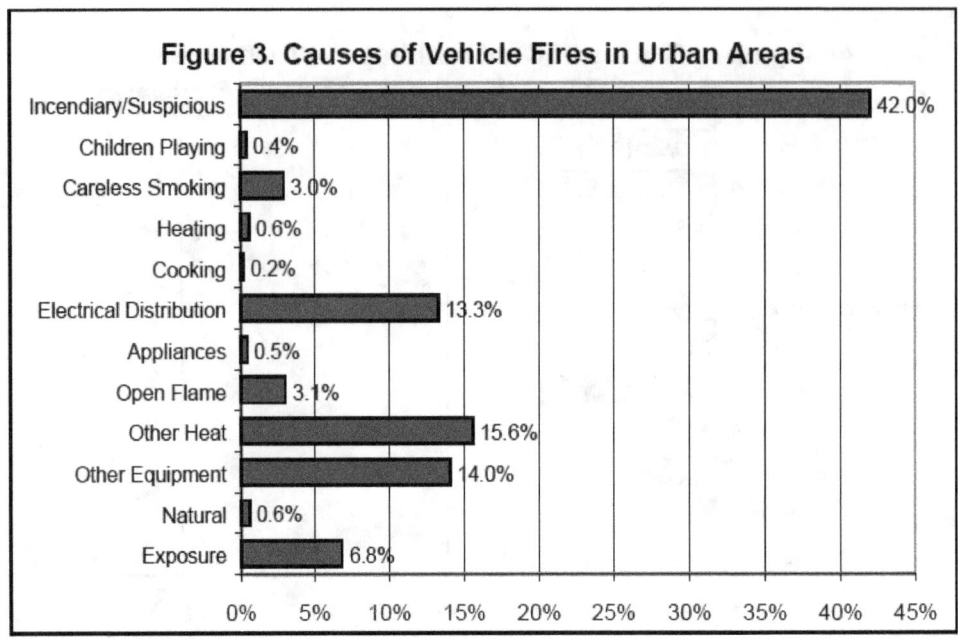

Source: 1996 NFIRS data

Unlike outdoor and vehicle fires, cooking is the leading cause of structure fires (Figure 4). Fires of incendiary or suspicious origin rank second, and heating third. The distribution of structure fires by cause is complicated by the diverse types of building occupancies. In particular it is useful to analyze residential and non-residential structures separately.

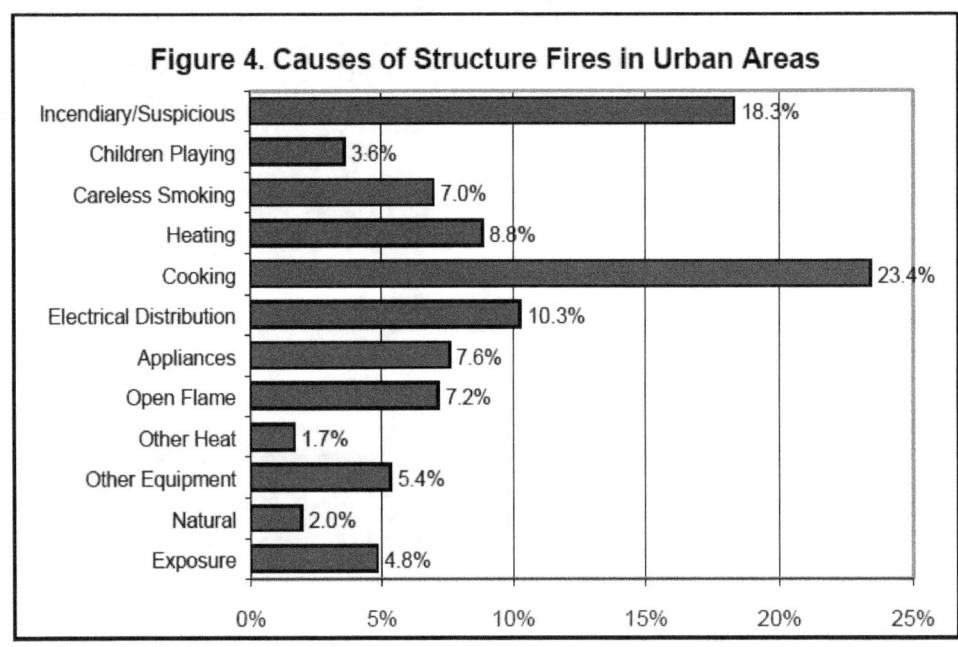

Source: 1996 NFIRS data

Structure Fires

Figure 5 provides a breakdown of structure fires by type of occupancy, whether residential or non-residential. Fires of incendiary or suspicious origin predominated among non-residential structures. This is an important observation as it indicates that a high proportion of the property loss due to fires in non-residential structures is avoidable. There is a broad range of motives for arson, including arson for profit; vandalism; spite or revenge; arson to conceal other crimes; or mental illness. Unfortunately, the current NFIRS system does not collect data on the suspected motive of arson fires. However, the next generation of NFIRS currently being piloted will include a module to collect this important information.

In residential structures, cooking, incendiary or suspicious origin, heating, and electrical distribution were the leading causes of fires. Cooking fires accounted for over one-quarter of all residential structure fires in urban areas. Incendiary or suspicious origin ranked second at 14 percent, while heating and electrical distribution tied for third at 10 percent.

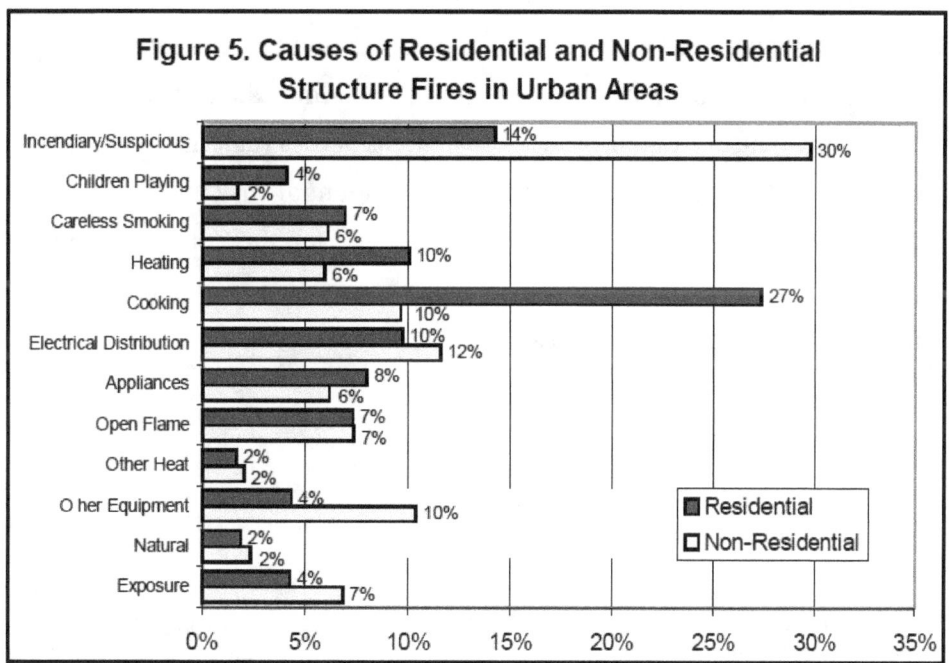

Source: 1996 NFIRS data

Residential Structure Fires

Since the majority of fire deaths and injuries each year result from fires in residential structures, the next several sections of this report deal specifically with characteristics of these fires. Expanding on the causes of residential structure fires, Table 2 breaks the U.S. down by region. It reveals that cooking was the leading cause in every region, and incendiary or suspicious origin was second in every region but the Northeast, where heating was second.

Table 2. Leading Causes of Urban Residential Structure Fires by Region, 1996

	Rank 1	Rank 2	Rank 3	Rank 4
Midwest	Cooking	Incendiary or suspicious origin	Smoking	Heating, Electrical distribution
Northeast	Cooking	Heating	Electrical distribution	Incendiary or suspicious origin
South	Cooking	Incendiary or suspicious origin	Electrical distribution	Heating, Open flame
West*	Cooking	Incendiary or suspicious origin	Open flame	Heating, Electrical distribution

* Estimated figure based on adjusted data. See footnote 7.
Source: 1996 NFIRS data

The leading causes of fatal fires tend to be different than the leading causes of all fires.[8] In particular, smoking is typically the nation's leading cause of fatal fires. Table 3 displays the leading causes of fatal fires in urban areas across regions. As with all fires, the table reveals that the top two leading causes of residential structure fires are fairly consistent across regions. In the Northeast, the Midwest, and the South, smoking was the leading cause of fatal fires. In the West, in contrast, incendiary or suspicious origin was the leading cause of fatal residential fires.

---

[7] In the West, some of the data for California were adjusted for conversion problems. Per discussions with California fire analysts, the leading cause of residential structure fires is cooking, yet for several fire departments the majority of these fires showed up in the "other equipment" category. Since the Los Angeles City Fire Department represented the vast majority of records with conversion problems, that city's data were adjusted to account for the conversion problem. Also, the Pacific Northwest was not represented in the analysis because there was insufficient NFIRS participation in 1996 from major metropolitan areas in that region.

[8] The same is true of the distribution of fire fatalities by cause of fire. For this analysis, it is preferable to use fatal fires as the unit of analysis because the distribution will not be skewed by instances in which more than one fatality resulted from a single fire attributable to one cause.

Table 3. Fatal Fires: Leading Causes of Urban Fatal Residential Structure Fires by Region, 1996

	Rank 1	Rank 2	Rank 3	Rank 4
Midwest	Smoking	Cooking, Children playing	–	Incendiary or suspicious origin
Northeast	Smoking	Incendiary or suspicious origin	Cooking	Open flame
South	Smoking	Incendiary or suspicious origin	Cooking, Electrical distribution	–
West*	Incendiary or suspicious origin	Cooking	Smoking	Electrical distribution

* Estimated figure based on adjusted data. See footnote 7.
Source: 1996 NFIRS data

Interestingly, heating does not appear as a leading cause of fatal urban fires in any region of the country. This is likely due to the widespread availability of central heating in urban areas where the climate is cold enough to warrant it. This is an important difference from the case of rural fires, where heating is the leading cause of fatal fires, followed by smoking.[9]

Type of Residence. One unique feature of urban fires is where they occur. Typically in the U.S., about 70 percent of residential structure fires occur in one- and two-family dwellings and 20 percent occur in apartments. In urban areas, however, apartments account for a higher proportion of fires. In 1996, 35 percent of urban home fires occurred in apartments, and 58 percent occurred in one- and two-family dwellings. This is not surprising given that apartments make up a higher proportion of the housing stock in urban areas than in rural areas.

Similar to all U.S. home fires, there are important differences in the causes of urban fires in one- and two-family homes versus apartments. Heating and electrical distribution account for more fires in one- and two-family dwellings than in apartments. In 1996, heating fires were over two and a half times more likely in one- and two-family homes (13 percent) than in apartments (5 percent). Similarly, electrical distribution fires occurred two times more often in one- and two-family dwellings (12 percent) than in apartments (6 percent).

Cooking fires, however, accounted for a significantly higher proportion of fires in apartments than in one- and two-family homes (39 percent versus 21 percent). With relatively low incidences of heating and electrical distribution fires, cooking as the leading cause of all residential fires in the U.S., takes on added prominence as the leading cause of apartment fires. This may, in part, reflect income differences between homeowners and renters. Gunther found that low-income groups experienced a higher rate of cooking fires than higher income groups, and the rate of poverty is higher among renters than homeowners in the U.S.[10] The distributions of urban residential structure fires by cause appear in Figure 6.

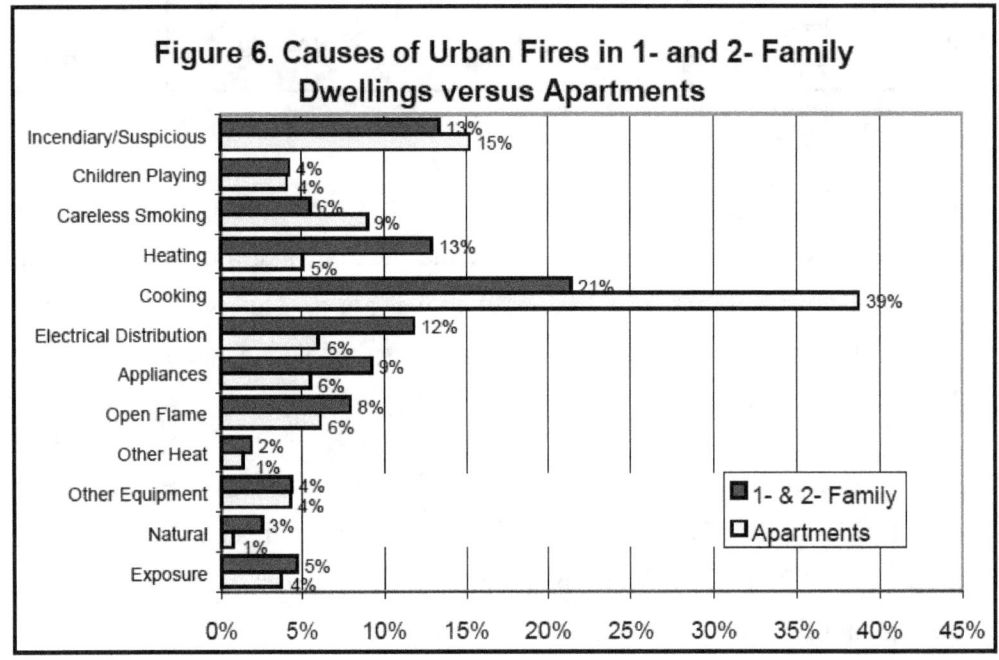

Figure 6. Causes of Urban Fires in 1- and 2- Family Dwellings versus Apartments

Source: 1996 NFIRS data

Month of Year. Home fires in urban areas are less sensitive to time of year than residential structure fires. This is largely due to the effect of heating fires, which accounts for fewer fires in urban areas than in rural areas. While more fires occur during the winter months than any other time of year, the difference between warm and cold months is modest (Figure 7). In contrast, a recent study of rural fires showed a more

[9] U.S. Fire Administration, 1998, *The Rural Fire Problem in the United States,* Publication FA-180/August 1998, (Emmitsburg, MD: U.S. Fire Administration), p. 16.

[10] Gunther, Paul, 1981, "Fire-Cause Patterns for Different Socioeconomic Neighborhoods in Toledo, Ohio," *Fire Journal,* (May), pp. 56-57.

dramatic difference in the number of fires from December through March compared to the warmer months.[11]

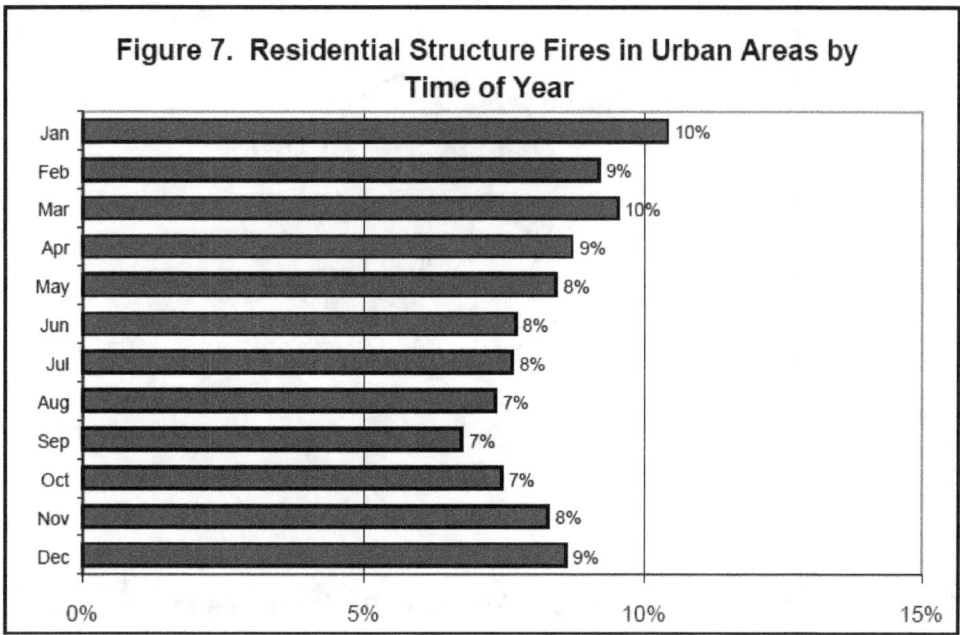

Figure 7. Residential Structure Fires in Urban Areas by Time of Year

Month	Percentage
Jan	10%
Feb	9%
Mar	10%
Apr	9%
May	8%
Jun	8%
Jul	8%
Aug	7%
Sep	7%
Oct	7%
Nov	8%
Dec	9%

Source: 1996 NFIRS data

Smoke Alarm Performance. Throughout the U.S., homes without working smoke alarms account for a disproportionate share of fires reported to local fire departments. Nationally, 28 percent of homes are not protected by working smoke alarms.[12] Typically, however, these homes account for about two-thirds of all residential fires each year.[13] Similarly, in 1996, 78 percent of fatal fires occurred in dwellings where smoke alarms were not present or were not in operational condition.

The same pattern applies in urban areas. Figures 8 and 9 present 1996 smoke alarm performance figures for urban fires. Figure 8 shows that 54 percent of urban homes experiencing fires that were reported in NFIRS did not have the protection of working smoke alarms. Similarly, Figure 9 shows that over two-thirds (69 percent) of urban fatal fires occurred in a home without a functioning smoke alarm.

---

[11] U.S. Fire Administration, 1998, *The Rural Fire Problem in the United States,* Publication FA-180/August 1998, (Emmitsburg, MD: U.S. Fire Administration), p. 24.

[12] U.S. Consumer Product Safety Commission, 1994, *Smoke Detector Operability Survey: Report on Findings (revised),* p. ii.

[13] U.S. Fire Administration, 1997, *Fire in the United States: 1985-1994,* Ninth Edition, Publication FA-173/July 1997, (Emmitsburg, MD: U.S. Fire Administration), p. 67.

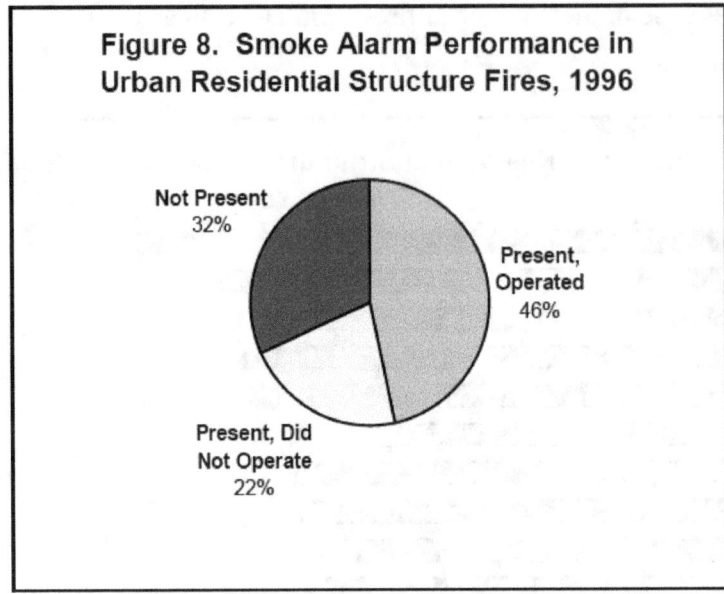

Source: 1996 NFIRS data

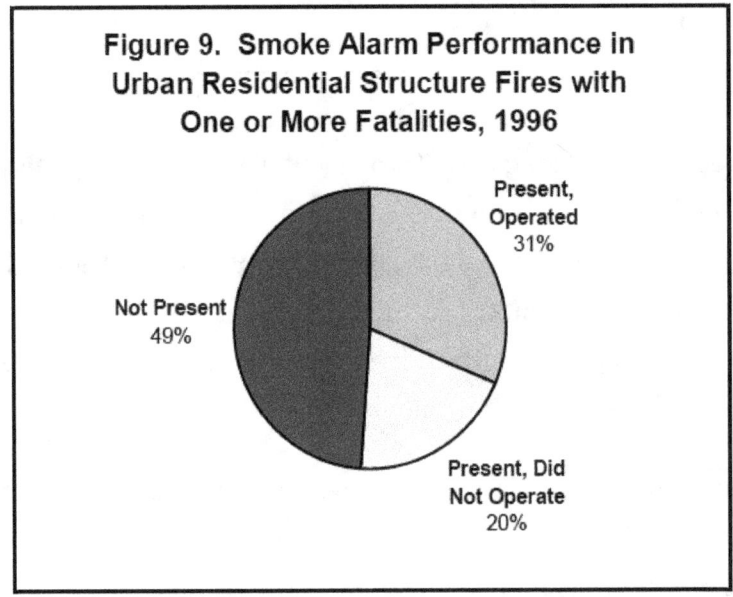

Source: 1996 NFIRS data

It is interesting that, compared to the U.S. as a whole, more urban fires occur in homes that have at least one functioning smoke alarm. Smoke alarms are intended to provide residents with early warning, ideally identifying a source of smoke before a fire starts. The increased incidence of fires in urban homes where a working smoke alarm is present is a serious cause for concern, and it should be the subject of future research. In the ninth edition of *Fire in the United States*, a similarly disturbing trend was noted. From 1990 to 1994, the proportion of homes with smoke alarms present that experienced fires rose by 33 percent.[14] In this study, homes experiencing fires in urban areas were more likely than the national average to have smoke alarms present and for those alarms to be operational.

Fire Death Rates

Mortality data from NCHS were used to calculate an overall fire death rate for the urban areas included in this analysis. All the fire death rates presented here are averages based on fire deaths occurring between 1991 and 1995. Together, the 18 metropolitan areas in this study had an overall fire death rate of 14.3 deaths per million population.[15] Expanding the data set to include the three metropolitan areas in the West (Seattle, Portland, and Phoenix) for which no NFIRS data were available, the urban fire death rate falls to 13.8 per million population.[16]

To put these rates into perspective, the fire death rate for the U.S. as a whole was 18.9 per million population. This provides indirect confirmation of the finding of a 1997 U.S. Fire Administration study that fire death rates were significantly higher in rural areas than in non-rural areas. In the present study, the urban fire death rate was lower than the overall U.S. rate due in large part to the fact that the U.S. rate captures the higher death rates in rural areas.[17]

Looking at fire death rates by region provides additional interesting information (Figure 10). For the 1991-1995 time period, the Midwest had the highest urban fire death rate, at 16.5 deaths per million population. The Northeast was second with a rate of 14.3, and the South was close behind at 13.8. The fire death rate was lowest in the West, where there were 10.0 fire deaths per million population. As demonstrated here, fire death rates tend to be lowest in regions of the country with the warmest and/or driest climates. This is an interesting finding and suggests support for recent USFA analyses that indicate climate and age of housing stock are strongly related to residential fire

---

[14] U.S. Fire Administration, 1997, *Fire in the United States: 1985-1994*, Ninth Edition, Publication FA-173/July 1997, (Emmitsburg, MD: U.S. Fire Administration), p. 67.

[15] This figure is based on total fire deaths in the urban counties with populations of one million or more included in this study.

[16] The inclusion of these cities makes the fire death rate presented here a more accurate estimate of the actual rate of fire deaths in all metropolitan areas in the West.

[17] U.S. Fire Administration, 1998, *The Rural Fire Problem in the United States*, Publication A-180/August 1998, (Emmitsburg, MD: U.S. Fire Administration), p. 38.

rates.[18] Cities with worse climates and older housing stock (here, the Midwest and Northeast) have a greater likelihood of fire and, potentially, fire death rates.

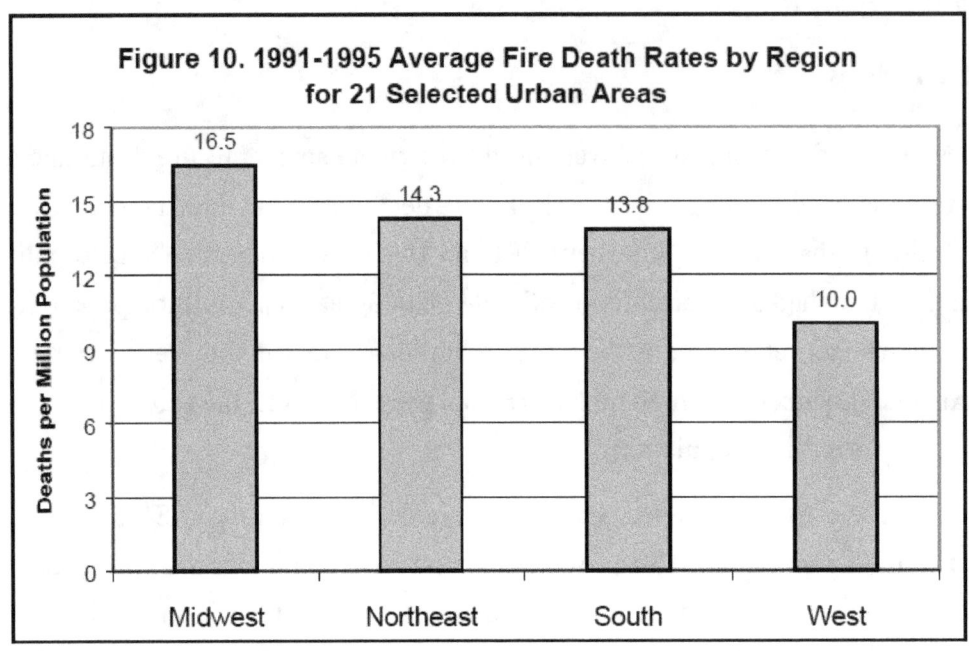

Source: 1996 NFIRS data

The final section of this report explores the fire problem in nine different metropolitan areas. These areas, as defined by the Census Bureau, are:

Atlanta
Boston-Worcester-Lawrence
Chicago-Gary-Kenosha
Cleveland-Akron
Dallas-Ft. Worth
Hartford
Houston-Galveston-Brazoria
San Francisco-Oakland-San Jose
Washington-Baltimore

Profiling these different areas can help identify similarities and differences among urban areas across the U.S. It should be noted that these profiles are based on NFIRS data reported from the participating fire departments in each metropolitan area. If those fire departments that do not report to NFIRS have fire problems vastly different than those that reported, this will not be reflected in the data presented below. To provide the fullest picture possible, all fire department and mortality data were included in the metropolitan

---

[18] U.S. Fire Administration, 1998, *An NFIRS Analysis: Investigating City Characteristics and Residential Fire Rates,* Publication FA-179/April 1998, (Emmitsburg, MD: U.S. Fire Administration), p. 2.

profiles – the analysis was not limited to those counties with populations of one million or more as in the analysis of the urban fire problem generally.

Metropolitan Profiles

DATA SOURCES AND NUMBERS

NFIRS, 1996

Fire Departments
  Reporting.................. 36
Fires...................... 10,034
Residential Structure
  Fires...................... 2,282

NCHS, Average 1991-1995

Fire Deaths Per Million
  Population.............. 12.7

SMOKE DETECTOR STATS

In 1996, smoke detectors were missing or not operational in:

- 53% of home fires
- 88% of fatal home fires

Atlanta MSA

1996 NFIRS Profile

Type of Situation Found. In 1996, 45 percent of all fires in the metro area were outdoor fires. Twenty-five percent were structure fires, 29 percent were vehicle fires, and one percent were "other" fires.

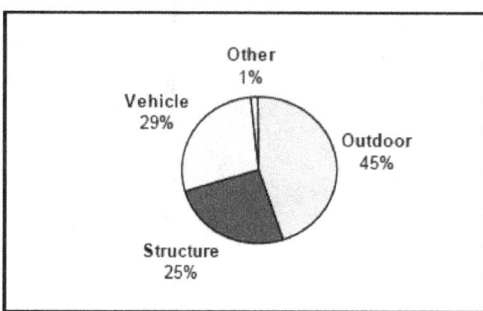

Causes of Structure Fires. The leading causes of residential structure fires were cooking, heating, and electrical distribution. In non-residential structure fires, the leading causes were incendiary or suspicious origins, open flame, and cooking.

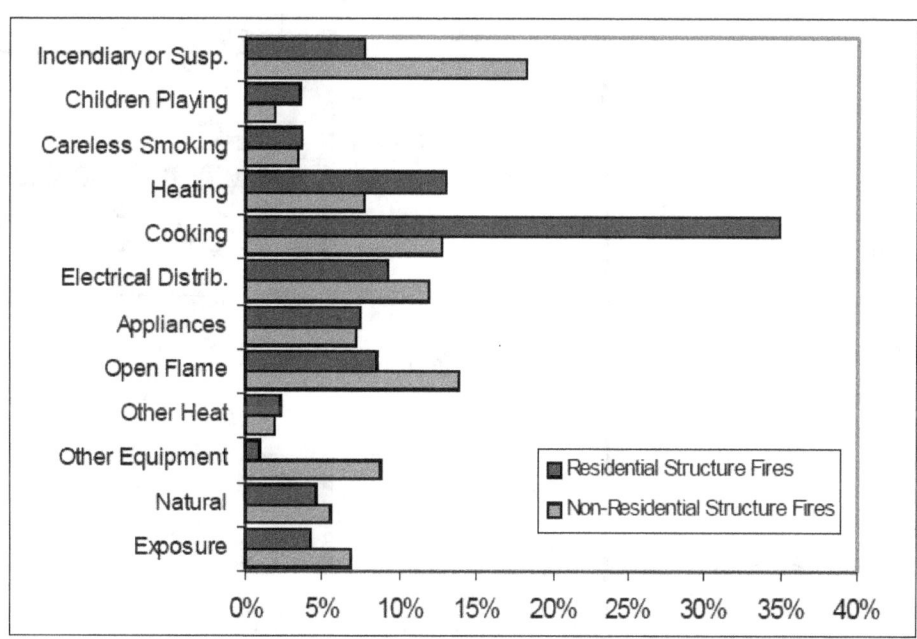

Causes of Residential Fatal Fires and Fires with Injuries. The leading causes of home fires that resulted in one or more fatalities were careless smoking, children playing, and electrical distribution. Of the fires in which one or more injuries were sustained, the leading causes were cooking, incendiary or suspicious origins,

	Cause 1	Cause 2	Cause 3
Fatal Fires	careless smoking	children playing	electrical distribution
Fires with Injuries	cooking	incendiary or suspicious	careless smoking

COUNTIES IN CMSA

Bristol County, MA
Essex County, MA
Hampden County, MA
Hillsborough County, NH
Merrimack County, NH
Middlesex County, MA
Norfolk County, MA
Plymouth County, MA
Rockingham County, NH
Strafford County, NH
Suffolk County, MA
Windham County, CT
Worcester County, MA

NOTE: Only Suffolk County is included in its entirety in the Boston CMSA. Census-defined portions of the other counties are included.

DATA SOURCES AND NUMBERS

NFIRS, 1996

Fire Departments
   Reporting................. 330
Fires....................... 28,246
Residential Structure
   Fires..................... 9,000

NCHS, Average 1991-1995
Fire Deaths Per Million
   Population.............. 11.5

SMOKE DETECTOR STATS

In 1996, smoke detectors were missing or not operational in:

● 38% of home fires
● 62% of fatal home fires

1996 NFIRS Profile

Type of Situation Found. In 1996, 29 percent of all fires in the metro area were outdoor fires. Forty-one percent were structure fires, 25 percent were vehicle fires, and five percent were "other" fires.

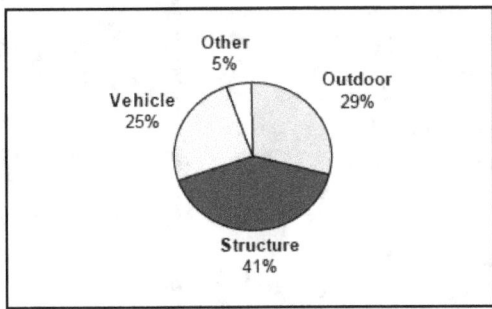

Causes of Structure Fires. The leading causes of residential structure fires were cooking, heating, and electrical distribution. In non-residential structure fires, the leading causes were incendiary or suspsicious origins, cooking, and electrical distribution.

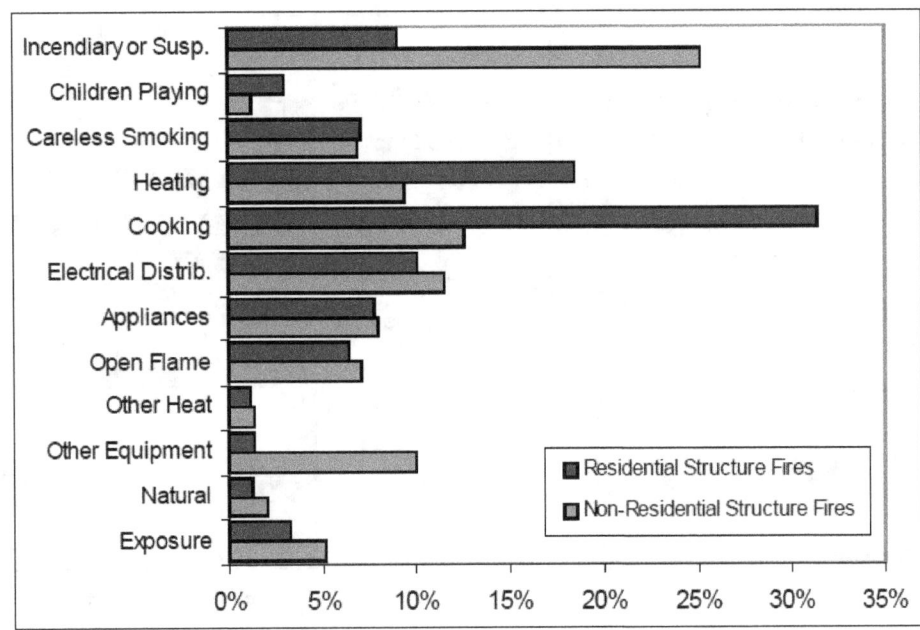

Causes of Residential Fatal Fires and Fires with Injuries. The leading causes of home fires that resulted in one or more fatalities were careless smoking, incendiary or suspicious origins, and cooking. Of the fires in which one or more injuries were sustained, the leading causes were cooking, careless smoking, and electrical distribution.

	Cause 1	Cause 2	Cause 3
Fatal Fires	careless smoking	incendiary or suspicious	cooking
Fires with Injuries	cooking	careless smoking	electrical distribution

Chicago CMSA

1996 NFIRS Profile

Type of Situation Found. In 1996, 44 percent of all fires in the metro area were outdoor fires. Twenty-six percent were structure fires, 28 percent were vehicle fires, and two percent were "other" fires.

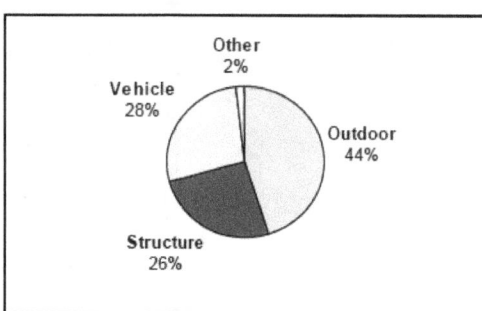

Causes of Structure Fires. The leading causes of residential structure fires were cooking, incendiary or suspicious origins, and careless smoking. In non-residential structure fires, the leading causes were incendiary or suspicious origins, exposure, and open flame.

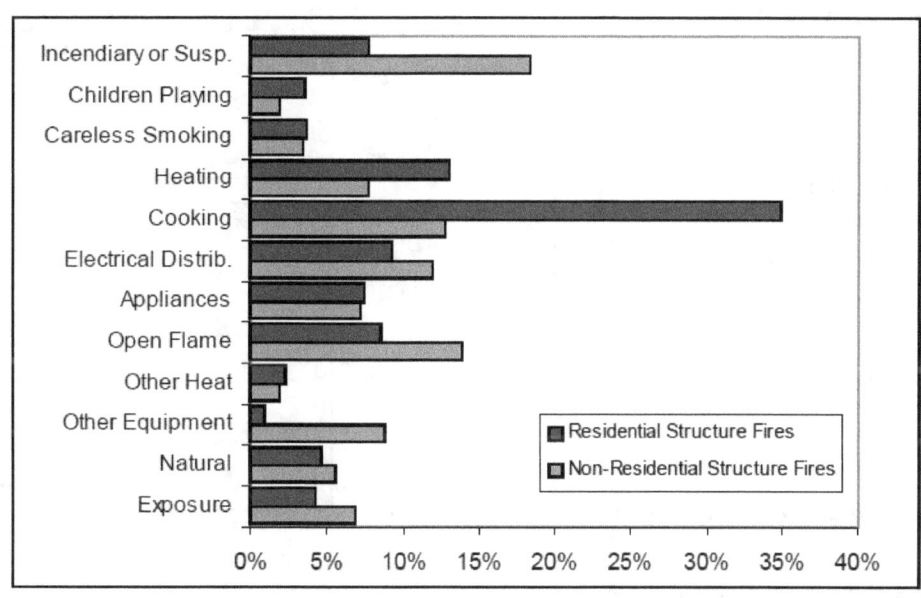

Causes of Residential Fatal Fires and Fires with Injuries. The leading causes of home fires that resulted in one or more fatalities were careless smoking, heating, and incendiary or suspicious origins. Of the fires in which one or more injuries were sustained, the leading causes were cooking, careless smoking, and incendiary or suspicious origins.

	Cause 1	Cause 2	Cause 3
Fatal Fires	careless smoking	heating	incendiary or suspicious
Fires with Injuries	cooking	careless smoking	incendiary or suspicious

COUNTIES IN CMSA

Ashtabula County, OH
Cuyahoga County, OH
Geauga County, OH
Lake County, OH
Lorain County, OH
Medina County, OH
Portage County, OH
Summit County, OH

DATA SOURCES AND NUMBERS

NFIRS, 1996

Fire Departments
  Reporting................ 142
Fires....................... 13,525
Residential Structure
  Fires..................... 3,371

NCHS, Average 1991-1995
Fire Deaths Per Million
  Population.............. 15.2

SMOKE DETEC- TOR STATS

In 1996, smoke detectors were missing or not operational in:

- 54% of home fires
- 58% of fatal home fires

Cleveland CMSA

1996 NFIRS Profile

Type of Situation Found. In 1996, 35 percent of all fires in the metro area were outdoor fires. Thirty-five percent were structure fires, 29 percent were vehicle fires, and one percent were "other" fires.

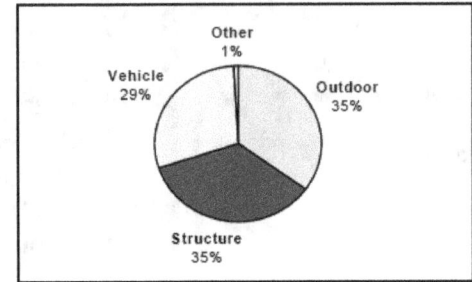

Causes of Structure Fires. The leading causes of residential structure fires were cooking, incendiary or suspicious origins, and heating. In non-residential structure fires, the leading causes were incendiary or suspicious origins, electrical distribution, and cooking.

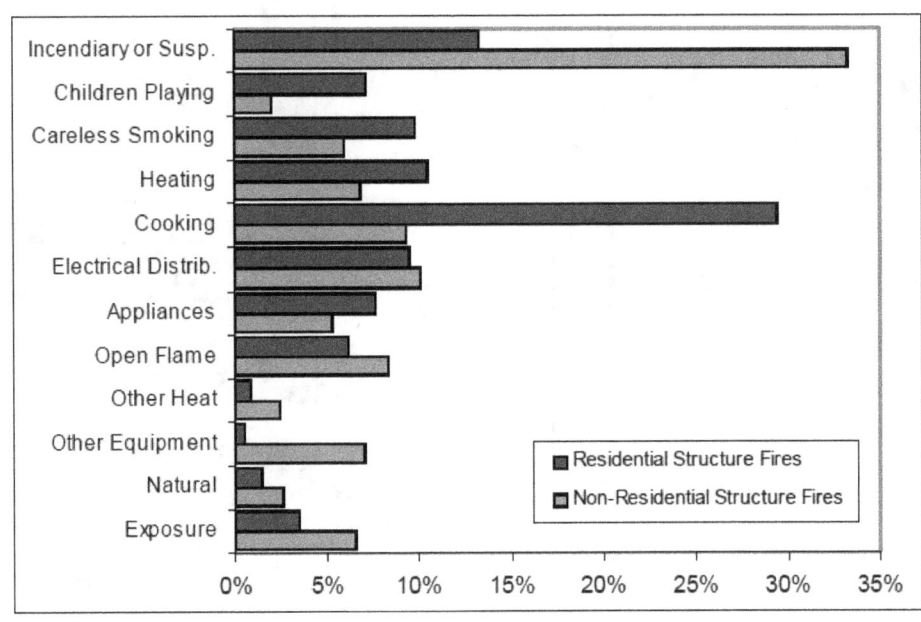

Causes of Residential Fatal Fires and Fires with Injuries. The leading causes of home fires that resulted in one or more fatalities were careless smoking, cooking, and children playing. Of the fires in which one or more injuries were sustained, the leading causes were cooking, careless smoking, and electrical distribution.

	Cause 1	Cause 2	Cause 3
Fatal Fires	careless smoking	cooking	children playing
Fires with Injuries	cooking	careless smoking	electrical distribution

COUNTIES IN CMSA

Collin County, TX
Dallas County, TX
Denton County, TX
Ellis County, TX
Henderson County, TX
Hood County, TX
Hunt County, TX
Johnson County, TX
Kaufman County, TX
Parker County, TX
Rockwall County, TX
Tarrant County, TX

DATA SOURCES AND NUMBERS

NFIRS, 1996

Fire Departments
Reporting 115
Fires 31,875
Residential Structure
Fires 5,787

NCHS, Average 1991-1995
Fire Deaths Per Million
Population 14.4

SMOKE DETECTOR STATS

In 1996, smoke detectors were missing or not operational in:

- 68% of home fires
- 69% of fatal home fires

1996 NFIRS Profile

Type of Situation Found. In 1996, 50 percent of all fires in the metro area were outdoor fires. Twenty-four percent were structure fires, another 24 percent were vehicle fires, and two percent were "other" fires.

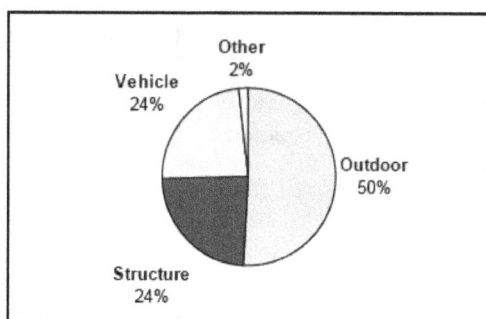

Causes of Structure Fires. The leading causes of residential structure fires were cooking, incendiary or suspicious origins, and heating and electrical distribution. In non-residential structure fires, the leading causes were incendiary or suspicious origins, electrical distribution, and cooking.

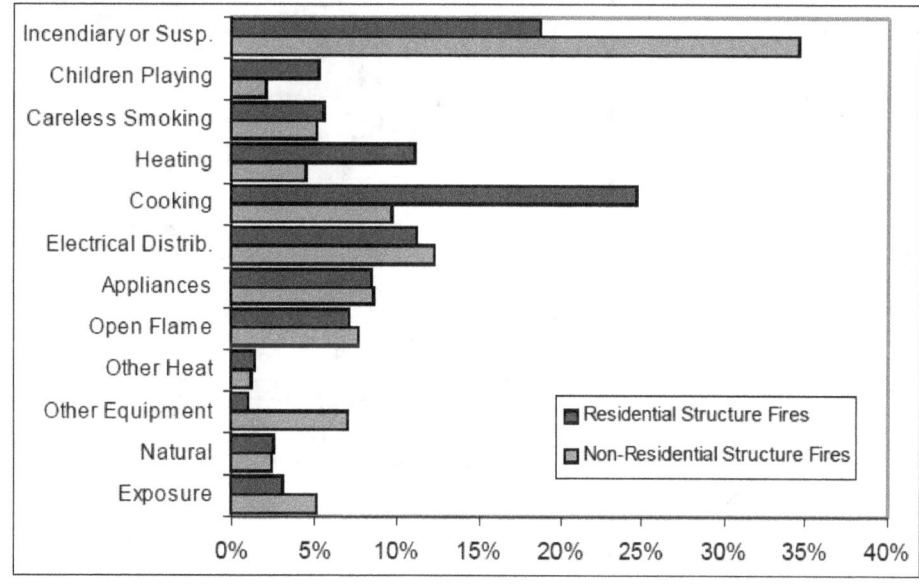

Causes of Residential Fatal Fires and Fires with Injuries. The leading causes of home fires that resulted in one or more fatalities were heating, careless smoking, and incendiary or suspicious origins. Of the fires in which one or

	Cause 1	Cause 2	Cause 3
Fatal Fires	heating	careless smoking	incendiary or suspicious
Fires with Injuries	cooking	electrical distribution	incendiary or suspicious

more injuries were sustained, the leading causes were cooking, electrical distribution, and incendiary or suspicious origins.

COUNTIES IN MSA

Hartford County, CT
Litchfield County, CT
Middlesex County, CT
New London County, CT
Tolland County, CT
Windham County, CT

NOTE: Only Windham County is included in its entirety in the Hartford MSA. Census-defined portions only of the other counties were included.

DATA SOURCES AND NUMBERS

NFIRS, 1996

Fire Departments
  Reporting................. 143
Fires......................... 5,998
Residential Structure
  Fires..................... 1,600

NCHS, Average 1991-1995

Fire Deaths Per Million
  Population................9.7

SMOKE DETECTOR STATS

In 1996, smoke detectors were missing or not operational in:

- 52% of home fires
- 63% of fatal home fires

Type of Situation Found. In 1996, 41 percent of all fires in the metro area were outdoor fires. Thirty-two percent were structure fires, 25 percent were vehicle fires, and two percent were "other" fires.

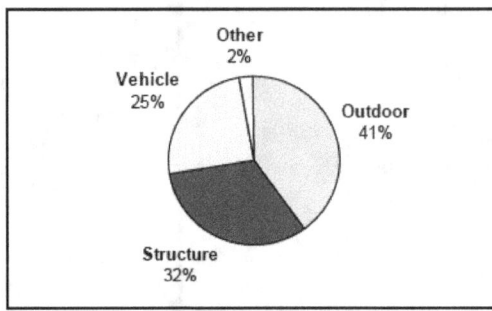

Causes of Structure Fires. The leading causes of residential structure fires were cooking, heating, and electrical distribution. In non-residential structure fires, the leading causes were cooking, incendiary or suspicious origins, and electrical distribution.

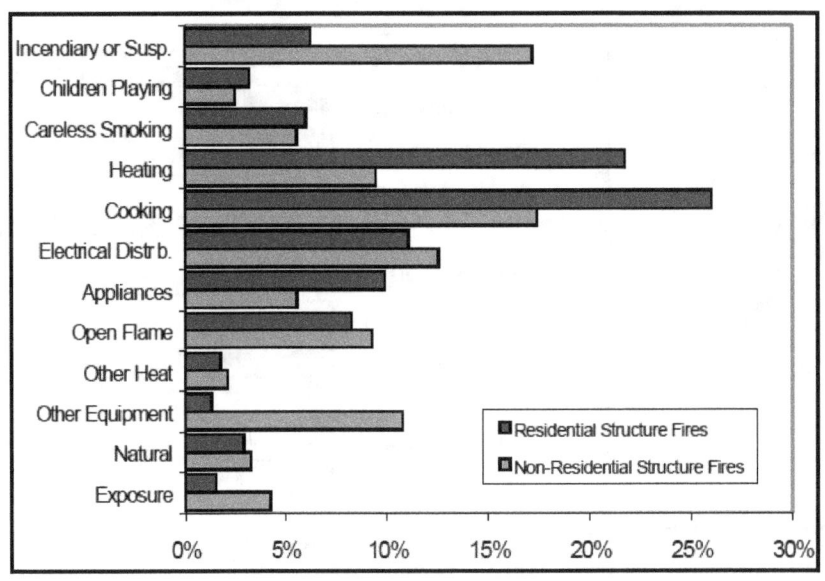

Causes of Residential Fatal Fires and Fires with Injuries. The leading causes of home fires that resulted in one or more fatalities were careless smoking and incendiary or suspicious origins. Of the fires in which one or more injuries were sustained, the leading causes were cooking, open flame, and careless smoking.

	Cause 1	Cause 2	Cause 3
Fatal Fires	careless smoking	incendiary or suspicious	5-way tie for third
Fires with Injuries	cooking	open flame	careless smoking

COUNTIES IN CMSA
Brazoria County, TX
Fort Bend County, TX
Galveston County, TX
Harris County, TX
Liberty County, TX
Montgomery County, TX
Waller County, TX

Houston CMSA

1996 NFIRS Profile

Type of Situation Found. In 1996, 44 percent of all fires in the metro area were outdoor fires. Thirty percent were structure fires, 25 percent were vehicle fires, and one percent were "other" fires.

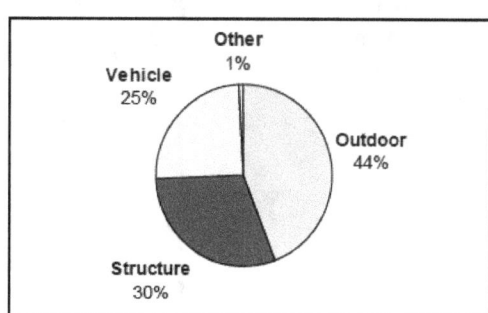

Causes of Structure Fires. The leading causes of residential structure fires were cooking, incendiary or suspicious origins, and electrical distribution. In non-residential structure fires, the leading causes were incendiary or suspicious origins, electrical distribution, and appliances.

DATA SOURCES AND NUMBERS

NFIRS, 1996

Fire Departments
   Reporting.................... 71
Fires........................ 19,742
Residential Structure
   Fires...................... 4,501

NCHS, Average 1991-1995

Fire Deaths Per Million
   Population.............. 16.0

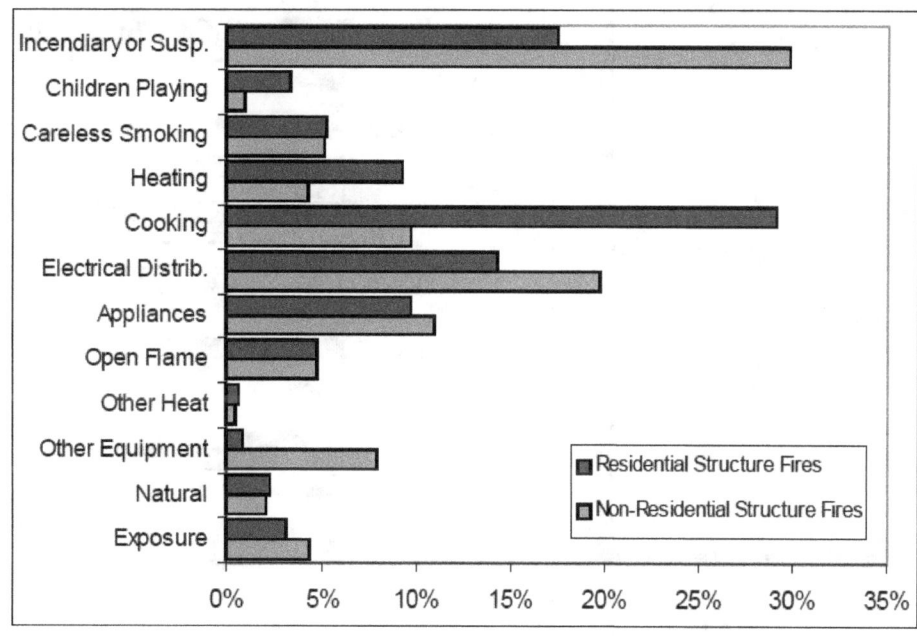

SMOKE DETECTOR STATS

In 1996, smoke detectors were missing or not operational in:
- 73% of home fires
- 100% of fatal home fires

Causes of Residential Fatal Fires and Fires with Injuries. The leading causes of home fires that resulted in one or more fatalities were incendiary or suspicious origins, careless smoking, and cooking. Of the fires in which one or more injuries were sustained, the leading causes were cooking, incendiary or suspicious origins, and electrical distribution.

	Cause 1	Cause 2	Cause 3
Fatal Fires	incendiary or suspicious	careless smoking	cooking
Fires with Injuries	cooking	incendiary or suspicious	electrical distribution

DATA SOURCES AND NUMBERS

NFIRS, 1996

Fire Departments
  Reporting................... 63
Fires........................ 11,894
Residential Structure
  Fires..................... 2,417

NCHS, Average 1991-1995
Fire Deaths Per Million
  Population.............. 11.9

SMOKE DETECTOR STATS

In the majority of cases, smoke detector status was listed as unknown for residential fires reported in NFIRS by San Francisco area fire departments.

San Francisco CMSA

1996 NFIRS Profile

Type of Situation Found. In 1996, 44 percent of all fires in the metro area were outdoor fires. Twenty-nine percent were structure fires, 25 percent were vehicle fires, and two percent were "other" fires.

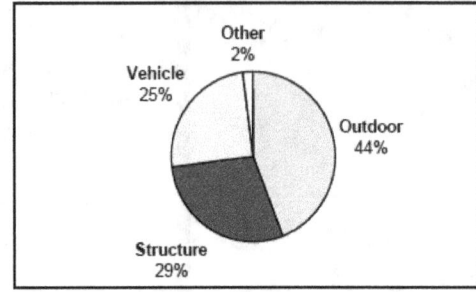

Causes of Structure Fires. The leading causes of residential structure fires were cooking, incendiary or suspicious origins, and open flame. In non-residential structure fires, the leading causes were incendiary or suspicious origins, other equipment, and cooking.

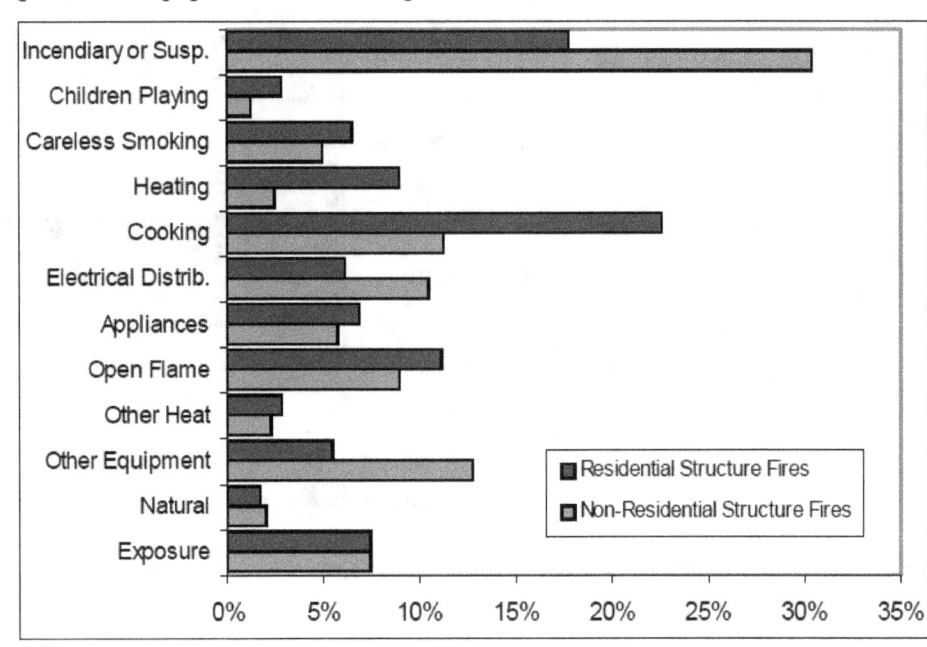

Causes of Residential Fatal Fires and Fires with Injuries. The leading causes of home fires that resulted in one or more fatalities were careless smoking, incendiary or suspicious origins, and cooking. Of the fires in which one or more injuries were sustained, the leading causes were cooking, incendiary or suspicious origins, and open flame.

	Cause 1	Cause 2	Cause 3
Fatal Fires	careless smoking	incendiary or suspicious	cooking
Fires with Injuries	cooking	incendiary or suspicious	open flame

DATA SOURCES AND NUMBERS

NFIRS, 1996

Fire Departments
  Reporting.................. 288
Fires...................... 32,433
Residential Structure
  Fires..................... 7,961

NCHS, Average 1991-1995
Fire Deaths Per Million
  Population.............. 14.7

SMOKE DETEC-TOR STATS

In 1996, smoke detectors were missing or not operational in:

- 52% of home fires
- 77% of fatal home fires

Washington-Baltimore CMSA

1996 NFIRS Profile

Type of Situation Found. In 1996, 41 percent of all fires in the metro area were outdoor fires. Twenty-nine percent were structure fires, 28 percent were vehicle fires, and two percent were "other" fires.

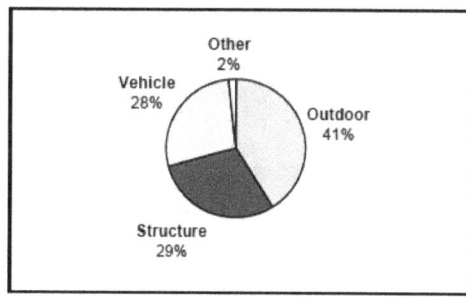

Causes of Structure Fires. The leading causes of residential structure fires were cooking, heating, and incendiary or suspicious origins. In non-residential structures, the leading causes of fire were incendiary or suspicious origins, electrical distribution, and cooking.

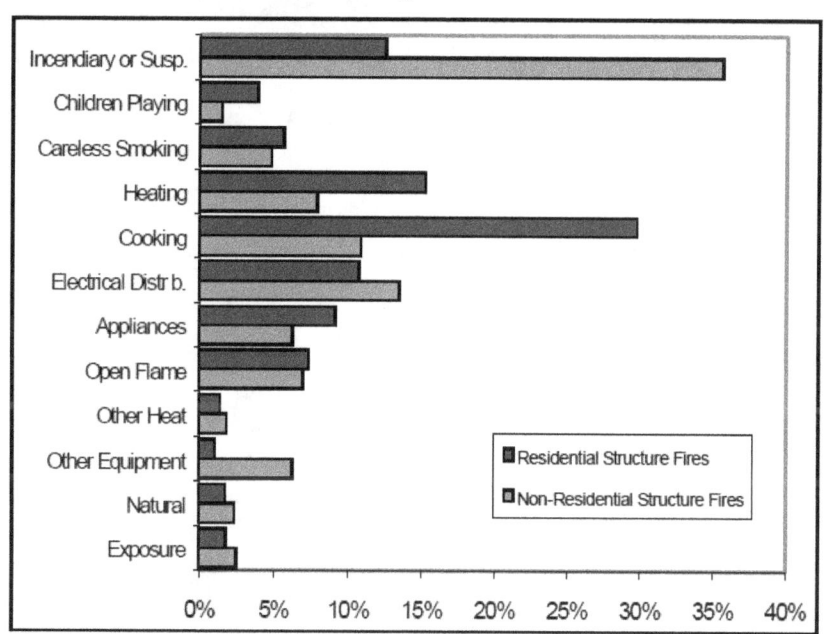

Causes of Residential Fatal Fires and Fires with Injuries. The leading causes of home fires that resulted in one or more fatalities were careless smoking, children playing, and electrical distribution. Of the fires in which one more injuries were sustained, the leading causes were cooking, incendiary or suspicious origins, and careless smoking.

	Cause 1	Cause 2	Cause 3
Fatal Fires	careless smoking	children playing	electrical distribution
Fires with Injuries	cooking	incendiary or suspicious	careless smoking